Where there is will there is a way

WHAT NOT TO DO WHILE **QSAR** MODELING

Authored By

Nivedita Chatterjee
Aayush Chowdhury
Sayantani Garai
Dipro Mukherjee
Pratim Chakraborti
Partha Sarathi Ghosh

Copyright © Nivedita Chatterjee. 2023
All rights reserved

This book is dedicated to all the novices in the field of Cheminformatics and computational biology.

This book questions all the most basic aspects of data handling and QSAR modeling and describes how to create a model properly while looking out for mistakes and how to efficiently perform error handling.

CONTENTS:

Chapters	Page
The need of understanding the QSAR Modeling	11
Data Selection: Process versus Data?	15
Mastering the art of selection of Descriptors (2D and 3D)	23
The process of model selection	29
Understanding the output of the model	35
How the Data and Model validation go Hand in Hand?	48
Deep understanding of R^2 and Rm^2 model validation scores- A low score isn't always bad?	52
Concluding Statements	56
Glossary	59

The need of understanding the QSAR Modeling

In Cheminformatics, there is an *in-silico* process called QSAR that chemists use to make predictions about how different chemicals/compounds will interact *with Cellular, environmental, and molecular components*. Some of these models work well and have been around for a while, but others are just simply wrong because model building is hard and they're always easy to mess up with the *fancy* computer programs that we use.

While designing drugs and modeling the fate of chemicals, QSAR models play an important role. These models connect chemical properties to biological activity and are used by regulatory agencies in making decisions about safety assessments.

There are two main types of QSAR model development: those that predict using regression analysis and those that use classification techniques.

The first and most important point to be noted here is that machines follow rules with efficiency, the very rules that we, the humans, set. So, the type of data we enter is the sole deciding factor for the outcome. But if you are not sure and enter values/ data with no proper relation or sequence, models will produce garbage even when it looks like things should be working fine. The application of this evolving technology has not only increased the efficiency but also enhanced the accuracy during various stages involved in drug discovery processes. This, therefore, has led to faster identification and development of novel drug candidates with higher therapeutic potential.

QSAR modelling has the following basic principles that are followed to validate a model:

    a.   Representation of Chemicals through 2D or 3D molecular descriptors.

    b.   Using statistical methods like Multiple Linear Regression (MLR) and stepwise Multiple Linear Regression (MLR) on the data set.

    c.   Validation of the model's stability through external as well as internal methods of validation.

    d.   Validating the predictability of the model

    e.   Interpretation of the final descriptors selected after using the model

A typical QSAR equation is elaborated as a Linear Equation, where:

Biological Activity $= Const + (C_1 \times P_1) + (C_2 \times P_2) + (C_3 \times P_3) + \ldots \ldots \ldots + (C_n \times P_n)$.

The descriptors of the molecules in the series are assigned parameters (P_1-P_n) and coefficients (C_1-C_n) and is based on their biological activity. Statistical techniques are used to derive a QSAR by analysing variations in these parameters. QSAR, being a very sensitive modelling technique, depends on the quality of the data and the choice of descriptors (what descriptors are would be explained in a later chapter).

This is what the book caters.

There are several books about how to do QSAR modeling with rules and different attributes that one needs to consider, all the correct processes one should proceed to undertake to create a good model and then execute model validation through different statistical validation matrices, just like we have explained earlier.

This book does not do that. This book doesn't speak of any of the ways a QSAR model should be created, validated and so on… This book speaks of all the *mistakes* one makes while going through the process. This book speaks of all the processes not to undertake while QSAR modeling, from the experience so far… Steps like Data Selection, Descriptor Selection, Scaffold Selection, Model selection and finally Model validation comprise the backbone of QSAR modeling and are the steps where recurrent errors happen, and that can completely change the outcome of on-going research.

We noticed that these mistakes are very hard to catch for beginners who are undertaking QSAR modeling; these mistakes are easily overlooked due to either a 'perfect' dataset or a 'perfect' score during validation. But what QSAR does is an in-depth analysis, where the aim needs to be very precise, that then slowly reveals the errors and problems after reviewing papers over and over again (we had to do the same 9 odd times). However, after finally locating the problem/s, the paper itself loses its value; completely. It shatters the trust upon understanding and the computational process.

To stop this from happening, to save time and to point out all the signs that the paper or the model might have a problem we are dedicating this book to all the novices in this field.

This book is a negative aspect-based and opinion-based guide to QSAR Modeling with real-time example, of our failed paper and why it failed step by step, so that it can come in handy to any newcomer in this field to look out for all the signs necessary.

Data Selection: Process versus Data?

Selection of data is the first step in QSAR modeling.

In the process of deriving an accurate QSAR model, meticulous selection of the representative compounds must be employed for preparing the data set. The inclusion of a larger number of compounds can significantly enhance model stability and improve its predictive performance. Therefore, great care needs be taken while ensuring that reproducible activity data from smooth dose-response curves (toxicity curves) with minimal errors are utilized to ensure reliability and quality in QSAR models.

Additionally in this regard, among the plethora of descriptors selected, free-energy related descriptors have emerged as valuable tools for enhancing the accuracy of QSAR studies. Thus, expressing biological activity in term of free energy (such as equilibrium or rate constants) has become pertinent when investigating the molecular properties, chiefly lipophilicity, or hydrogen bonding capacity.

Now, let us understand how exactly data is selected for studies of these kinds.

The following are the aspects of selecting data for QSAR:

1. Data/ molecules selected due to structural similarity
2. Selection of data due to scaffold similarity
3. Selection of data due to them belonging to the same class or family of molecules

Note a 4th aspect can be selection of data depending on a specific disease, and then going by class of molecules.*

What we did in our paper, is that we selected our molecules based on structural similarity. Now, this process involves pinpointing molecules that have very similar structures, even though they are of different class or have different biological activity. We selected our molecules, a set of 56, from a Data set of 1200 molecules.

Figure: The figure shows some structures of the molecules that we selected for this study; Observe how they do not exhibit a proper backbone structure except a 'phenol' that is common for all molecules, this is because we tried selecting all double benzene ring structures before even studying the term 'scaffold'.

Our dataset was provided during a national hack-a-thon therefore, the adequate amount of data (for the time when we had selected our data set). But then again we took several steps only to understand that the data itself was only a one time testable data, with a lot of loop holes that we had missed or overlooked.

Let's start with our first mistake...

We based our dataset of 56 molecules solely on a manual search of structural similarity. We ran the main molecule's SMILES (Simple Molecular Input and Line Entry System) through the PubChem fingerprint algorithm to confirm the data set and just set off to do QSAR Modeling.

Second mistake...

We did a manual search, meaning we ran several SMILES through the SWISSADME application to understand and understood their structural similarity and made our choice. This led to two problems:

    a. We had no way of knowing their biological activity scores/ similarity

    b. We proceeded with this data even though they were vastly different from each other in terms of class.

But the most integral step that we overlooked was that these molecules did not have any defined background of clinical trials.

So, here is the thing, when we do undertake something like drug-likeliness, or toxicity tests through QSAR we need to keep one thing in mind, this process is only feasible if there is ample valid literature that can state that the chosen molecules for a model are actually fit or not fit to be a drug.

This is the reason why during this process, researchers usually take drug molecules that have failed during clinical testing either due to a functional group or due to an estranged scaffold that seemed like an efficient drug compound but had too many side effects and so on…

Now, this is where our next mistake needs to be addressed… 'SCAFFOLDS'

What most novices do while writing their first QSAR research is: Select data, run that data through a model, miraculously get a high score during validation and end it there. But then their study gets rejected during publication.

Why does this happen?

This is because the basis of their data selection doesn't even exist. This basis is what a **scaffold** creates because a Scaffold describes something called a unique structure. Molecular 'scaffolds' are crucial building blocks in various projects that involve small molecules, such as drug design and discovery.

It is used for identifying core structures which serves as the foundation or initial point of synthesis or structure optimization. Two of the Widely accepted definitions include Bemis Murco scaffolds and analog series-based scaffolds. These frameworks have proven quite effective in extracting systematic information from several compounds with regard to chemical compositions/ similarity.

The process of identifying classes of chemically and biologically similar compounds is non-trivial that involves careful consideration of various attributes. At the heart of this process lies the need to ensure that all compounds within a given class share structural similarities and therefore mode of action, while remaining distinct enough to elicit systematic changes in biological activity. For deriving the similarity attributes such as backbone structure, substituent, reactivity patterns, a molecule's biological activity and/ or toxicity values can be considered.

What might happen during data selection, therefore, is that the molecules can be structurally sound and even similar looking, but they might not share a common scaffold. This in turn gives rise to an increase in number of outliers in a study, not because they are unique but because they are actually a completely different class of molecules.

Going ahead with this, it leads to a data set of molecules with no defined backbone structure and therefore, taking one further away from concluding anything, at all whatsoever about the nature of these molecules.

Now, for the last mistake that one can make during this process of data selection, it speaks of the aim. Yes, you read it right, the aim of the study.

Usually, QSAR is either done on environmental/ disease specific drug toxicity or to understand drug-likeliness of a disease-specific compound. Our work falls in the second category where we had selected drug-like compounds that proved to be permeable when administered on the Caco2 cell line; the keyword being 'Disease-specific' for both cases. Our aim for the paper was to understand if the selected molecules were effective against intestinal cancer, while the data / molecules that we had only had literature on their permeability on the intestinal cell line. We did find some studies that spoke of intestinal/ colon cancer, but they came out to be contradictory.

Our mistake was that we were not aware that just because a drug-like compound is permeable o a specific cell line, it doesn't establish the fact that they are effective, which many novices (including us) end up concluding. Therefore, aim of the study needs to be crystal clear before selecting any compound at all.

This is called Data filtering, because this is where we create a set of meaningful data that are related to each other and are specific to a cause. This is the reason why we need to be extra cautious during data selection and paper designing.

Therefore, the following points are to be kept in mind:

1. Do not make the mistake of selecting similar looking compounds as being the same class of compounds
2. Never overlook the main backbone structure of a molecule, termed as molecular scaffold.
3. Do not willfully select any molecule that fits the problem statement, no matter from which data set it is;
4. Always try to go for disease specific compounds/ reason-specific compounds, while conducting this study.

5. Never overlook the aim, just because you feel like enough data is present; perform proper literature survey to understand your compounds and then frame a QSAR paper.
6. A good score doesn't always mean good data; therefore, try as many methods possible to recheck the model and its validation.
7. Please keep in mind – a low score doesn't mean the data is garbage.

Mastering the art of selection of Descriptors (2D and 3D)

Before we start to understand this section, let us discuss, what exactly a Descriptor is… and why this particular term is of such importance.

Descriptors or rather molecular descriptors have a role in elucidating the molecular information in Cheminformatics and preclinical Drug Discovery. Using mathematical algorithms, they provide precise numerical values to describe vital physicochemical properties of chemical compounds to have a deeper understanding of structure-activity relationship (SAR). QSAR practitioners can utilize these highly informative descriptors to evaluate various attributes related to reactivity, solubility or absorption that are not readily observable experimentally, aiding innovation in the field of research in biological systems.

As we progress further to understand, the term 'Descriptor selection', this itself becomes the biggest source of confusion that can be identified for any newbie experimenting QSAR and subsequently writing the article on QSAR.

This is because: we DO NOT SELECT.

The process of Descriptor selection is in itself a part of the study when the model would itself choose the descriptors and not something the researcher has the liberty to choose. This is why now we come to the most integral part of this process- understanding: -which descriptor is a dependent descriptor and which is an independent descriptor.

When it comes to a 'dependant descriptor' or rather variable/ attribute, we speak of a descriptor that describes a characteristic feature of a molecule directly like, permeability, reactivity, drug-likeliness (this is a disease specific feature) or indirectly and that too with a solid foundation like the number of hydrogen bonds present can describe the solubility and so on. For example, we can take a solubility descriptor (2D or 3D) like LogS, or we can take a descriptor that defines solubility in a physical manner, like a descriptor that talks of how many free Hydrogen are present, as we all know the greater number of free hydrogens, the easier it is for that molecule to be soluble; descriptors like 'Hydrogen bond count' speak of the same.

Now comes our 'independent descriptors/ variables/ attributes' which are characteristic features that simply describe a molecule physiologically and biologically. There are several thousands of independent variables that can describe (benzene) ring count, a molecule's toxicity through even more sub-divisions (e.g., *like organ toxicity, toxicity endpoints, toxicity pathways and toxicity stress response pathways*), a molecule's solubility and so on. These variables are not essentially related to one another and yet still define a molecule, independently.

During QSAR we study which descriptor is best suited to define out chosen set of molecules. This is when an algorithm chooses a set of descriptors that align or correlate best with the dependent variable or descriptor that the user provides.

For example, our dependant descriptor was Log P, a permeability descriptor, through which the model chose five descriptors, which describes the correlation to LogP and defines Log P, itself.

This then finally brings us to a point when the pandora's box of errors is unlocked:

1. We chose our own descriptors
2. We went to several different web servers to generate our descriptors
3. Even after getting a final set of descriptors generated by a server, we concatenated additional descriptors to the dataset.
4. The best descriptors set chosen by the model can describe both the correlation with the dependant variable as well as the toxicity of the data set (didn't pay heed to this rule when we did the study).

Note* *point 3 and 4 are simultaneous but have been written separately for defining the issue of the specific situation.*

Let us start with rectifying point 1...

One cannot choose manually their descriptors, as we do not have the precision or sound enough calculation to select a descriptor without loopholes, like-

➔ There might have been a descriptor that could completely validate the model

➔ The descriptors we select might have a good score but also have more mean square error

➔ They might not even be the best descriptors for that specific attribute

This can lead to problems such as a very haphazard score from the model, nearly no correlation between the descriptors and the worst-case scenario can be an over-fitted model which would give the most perfect score ever but in reality, isn't even a least feasible outcome.

In our case, during model validation we got a score of 0.992, with three outliers. This is something which made us insanely elated. We got the perfect regression plot that was overlapping the actual regression line; there were six descriptors that were perfectly correlated to our dependant variable.

After trying to verify the received score, we *accidentally* realized that there was an issue with our model. We learned about 'co-linearity' and discovered that our model had too many interdependent factors, leading to over-fitting. It turns out that using Multiple Linear Regression to define our model wasn't the best idea from the beginning. We realized this after getting rejected for the first time as we had not taken into account the variance inflation factor (VIF) of the model that clearly showed co-linearity among the set of descriptors. Usually what happens is that we see a good p-value i.e., less than 0.005 and we accept that the model is excellent, whereas a VIF score more than 1.5 is unacceptable and is possible even after p-value/ confidence level is 0.0005.

This points to two postulates:

    a.   The descriptors do define the dependant variable perfectly

    b.   This happens not due to them being independent variables but different variations of the same dependant variable.

For example, our dependent variable being LogP, got a set of best six variables containing three variations of LogP, namely XLogP, MLogP and SilicosITLogP. We even got the perfect QSAR linear equation only to reject it later.

This happened due to point 2…

We selected our descriptors from not only one but multiple servers like SWIDDADME and Protox. This led to the problem where we didn't leave room open for any possible better descriptors that could have defined the model. We ended up explicitly choosing only the attributes we saw befitting and added that data into the model… and hence, no wonder the model was over-fitted.

This wasn't the final mistake though, since now we come to point 3 as well as 4…

We didn't have any idea that even though an attribute may (for example) define solubility, if chosen by the model, we must never try to add any more attributes for defining, say, probable toxicity of our selected molecules.

When a model selects an independent descriptor based on the dependant variable, it already established that the descriptor is the most feasible and efficient attribute to define that quality (say permeability) of the molecules. Adding more descriptors just because we think the chosen final descriptors are not adequate, that too doing so manually, is nothing but adding plethora of errors to the process.

Now, after finding the errors that manual descriptor selection can add to a model, we ended up taking a descriptor set of 1545 from PaDEL, where we generated 2D descriptors and let the model decide which descriptors to retain and which to reject. This led to two conclusive opinions:

    a.   We had a set of five best descriptors that were linear.

    b.   These descriptors were not over-fitted and still portrayed high values of correlation.

Our ANOVA table now exhibited all the values that pointed towards a linear model and that validated our model with a score of 0.871. But this time, the number of outliers had increased, and they showed some variation in terms of biological activities as this final set of descriptors were completely different from what we had selected in the past from SWIDDADME. These descriptors specifically defined the permeability of all the molecules and not anything else.

This is where we finally realized that after performing an actual descriptor selection, the very aim of our paper was in question. This happened significantly due to two reasons:

1. Lack of initial investigation to prove any kind of relation of LogP to drug-likeliness
2. Lack of literature survey to find actual papers and not reports to prove clinical feasibility of any drug-like characteristic at all.

This is one of the most important reasons, why selection of descriptors may seem to be the easiest process ever but is like a ship treading through murky waters. The most significant takeaway points to be remembered from this chapter are that:

1. Do not select descriptors manually, strictly no.
2. Do not select descriptors from various web servers, select any one and follow that holy grail.
3. Let the model select best fit descriptors in terms of correlation.
4. Pay close attention to if the chosen final descriptors are co-linear or not
5. A great validation score does not mean the model is correct.
6. Understanding the Dependant variable/ descriptor is the most integral part of model building.

The process of model selection

Absence of Evidence is not evidence of absence

When it comes to model selection during QSAR, we need to pay close attention to the prior two chapters as they precisely decide what model one would need to apply for their specific research.

Creating a mathematical equation that relates physiochemical parameters to biological reactions, explaining all molecules in the dataset is important. Relevant independent variables, known as molecular descriptors, need to be identified while removing/ changing any potential anomalies in the data. **This is what a Model does.**

The following steps are followed during selection of a model based on number of data present:

1. If the data set is more than 2000, we usually perform Random Forest Regression
2. If the data set is smaller than 2000, we usually employ Multiple Linear Regression

For QSAR, one usually employs multiple linear regression (MLR), since data sets usually range from 55- 1000, which is a very small range, and the most important point to be kept in mind is that their descriptors are assumed to be **linear** and **independent** while one employs a model to use that data.

One dependent descriptor is always present and that decides the fate of the other descriptors and the dataset, for us this dependent variable was LogP. Now, when we do a QSAR study, to scale down the number of descriptors or filter the descriptors that best correlate with the dependent variable, we do Stepwise MLR. It can be either forward or backward stepwise depending on the number as well as nature of descriptors.

The whole process is as follows:

1. Feature selection: To select best features with less error
2. Feature significance: to select best features with more correlation
3. RMSE (Root Mean Square Error), MSE (Mean Square Error), MAE (Mean Absolute Error)
4. R^2 and Rm^2 for validation and regression plot generation

It is during Feature selection and Feature significance that the model gives us the best descriptors as output in form of a graph, or even a heat-map to show correlation scores of independent variables with the dependent variable. This is the step of descriptor selection by using MLR.

Now, we employ the forward or backward stepwise selection of feature, by filtering out descriptors based on the RMSE, MSE and MAE values and re-confirming the descriptor set.

Through these processes we understand which data set and their descriptors snugly fits the aim of the research. For our work we used the forward stepwise selection and did 10-fold cross-validation and we got the best 5 descriptors from this.

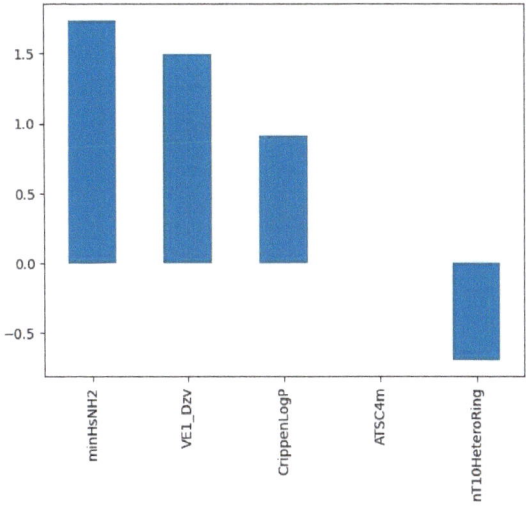

Figure 3.1: Feature significance through RMSE; error <0.0005 (correct descriptor set)

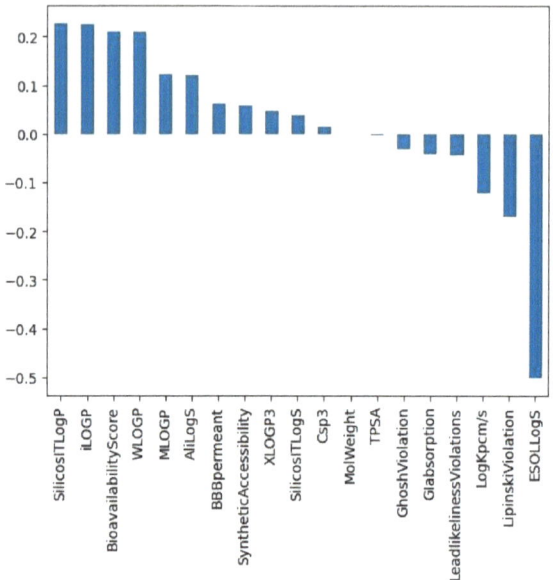

Figure 3.1: Feature significance through RMSE; error <0.0005 (incorrect descriptor set); observe how many variations of LogP are here

Now now... Why only MLR? You might be asking why we only use MLR for QSAR studies... Well, we don't.

We can also employ Partial Least Square (PLS) forward and backward stepwise if we specifically want to study collinear descriptors! This is usually done on datasets with a maximum range of 200 molecules, to understand collinearity between the dependent variable and so-called independent variable.

But the most important point of Model building as well choosing is understanding our data and descriptor set.

This is where the following mistakes happen:

1. Not knowing the nature of descriptors: whether they are linear or collinear

2. Not understanding molecule classes/ substructures

3. Taking molecules that are similar but have vastly differing descriptor values

Note* In our case we did all three of these mistakes.

The first error can miraculously give excellent output scores and yet the model will be completely wrong, this happens as collinear data is also redundant in some cases. The model thus gets over-fitted.

Usually, a model's nature during QSAR is to search unique descriptor data, which leads to proper understanding of a molecule, but only linear data does this, through which descriptors are filtered. **If there is any form of redundancy**, that descriptor is not selected any longer. And employing a linear method on colinear data would obviously yield good results due to them being connected to the dependent variable already. This mainly takes place as the person playing with the data doesn't not understand how a model needs to be employed based on the nature of the data, which is hard, but with ample literature review is possible.

The best way to do this though, is to employ both linear and colinear methods, and understand from scores of the ANOVA table, that which method needs to be employed, although that is not the actual process.

This is what we ended up doing due to our model being overfitted.

Employing this process would mean employing elimination method, which is completely unethical as that would mean we have tampered with the data depending on the outcome. This is not acceptable while doing QSAR which is why we must lay utmost emphasis on understanding the data and their descriptors.

Not just this, if we have taken an improper data set, but good descriptors, then the scoring matrix can give scores as low as -120.98355. Yes, we got this score after finally getting a proper set of descriptors, our model validation score went into negative.

At the end, the regression plot generated was something like a maze instead of a straight line, which proved our data was not correctly chosen or were filled with garbage values.

Therefore, during model selection, these salient things have to be ingrained in the process and followed:

1. Do not overlook the trend of the data.
2. Do not overlook the values of the descriptors.
3. Understand the nature of descriptors from linear to collinear.
4. Do not use elimination method: that is data manipulation.
5. Employ a linear model for linear data and a collinear model for collinear data.
6. Do not interpret a dataset as bad data due to a lower score.
7. Observe all descriptors and study about them beforehand, before employing a model.

Always remember, data never lies neither does a model; but loopholes do.

Understanding the output of the model

Clarity is a consequence of handling your confusion consciously

Now that we have understood how important the choice of a model is, let us embark to understand how to interpret the output.

The output of QSAR is very straight forward, thankfully, for the newbies… which makes it even harder to understand. Well, to start with, there are two points to be noted here:

1. If you are performing a QSAR alone without help from all the proper relevant references, it is often hard to grasp the meaning of the exact output.

2. If you're executing a model with your team, then most likely you can end up neglecting signs of the lethal errors that can ruin the whole QSAR process, without anyone's understanding, even if you get a good score.

Therefore, let us dig deep into how a QSAR output looks like, as we would bring about all the errors that can ruin the process. A model gives output in three forms:

1. Graphical Output
2. Heat-maps, and
3. An array of Score

Output 1:

Graphs: The graphical portrayal of molecular data using the final descriptor set.

A graph in a QSAR Model usually defines the dependent variable and the predicted dependent variable, for our case this was LogP and Predicted LogP respectively. These were predicted using the independent descriptors, that was selected by the model.

But again, if there is an error, the resulting graphs may fall into the following categories:

a. Over-fitted
b. Under-fitted
c. Completely haphazard (*failed model*)
d. Good fit

An over-fitted graph shows a beautiful overlap of the original regression line and the trend line. Here the molecules lie over the line showing that the model is at near perfection.

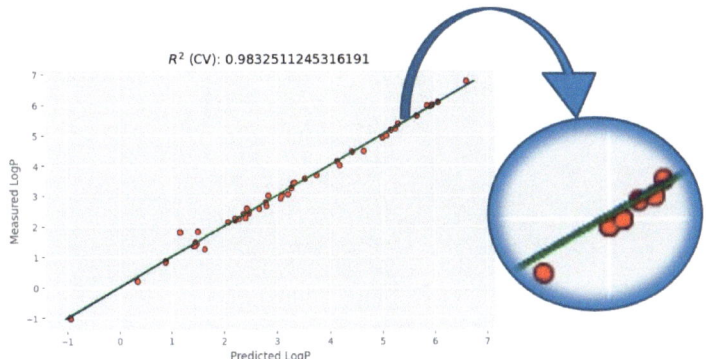

Figure1.a: How an over-fitted model's regression line is overlapped. (Green overlapping blue)

This was our first attempted QSAR where we had gotten a score of 0.99 for Rm^2 and 0.98 for R^2 validation. This led us to believe that the data set was extremely sequential, and these molecules had great permeability less toxicity as well.

But just because these molecules were permeable never meant that they had drug-like qualities, which was proved when we went ahead to further analyze their toxicity through other webservers.

It was found that most of the molecules in our data set fell under class I or II of toxicity with LD50 values of less than 30. This is **Lethal** for any human and can cause potential critical end-point due to blood toxicity, if ingested.

Therefore, a great Rm^2 value can never define the toxicity of the data set if there is no proper data set to begin with or proper descriptors to describe the properties. Either way, the results can be severely harmful in both theory and in-silico testing.

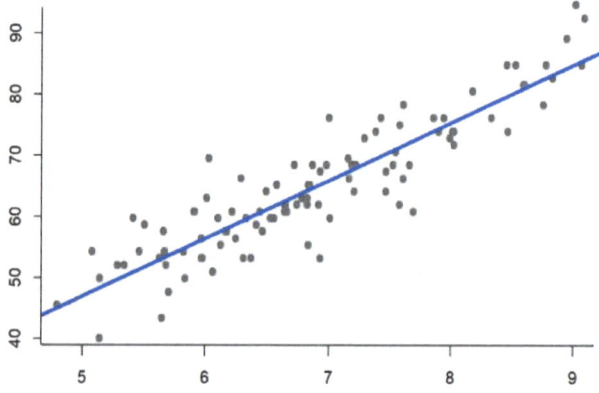

Figure 1.b: An underfitted Graph- observe how there are several outliers among such a small number of molecules and the regression line doesn't define the scores properly either.

The second type is the under fitted graph. This comes as an output if there is not enough data available. The graph to any third person would look normal but since the model can't make a proper analysis due to too less data, too many outliers are lined out, since the model isn't capable of understanding all the trends in the values. This is probably due to a shortage of enough proper or rather significant values. The minimum data set needed for a QSAR model to work is 55or above whereas the maximum is 2000. This condition is for small data sets, where data set defines a number of molecules. When it comes to large datasets, we mostly use Random Forest Regression to define our data. The plots received during these are analyzed similarly.

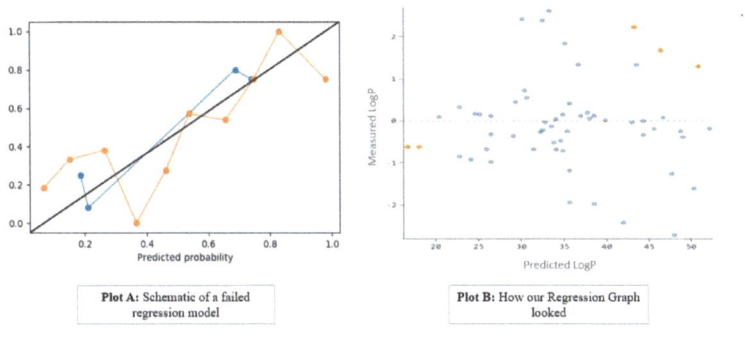

Figure 1.c: The figure shows what can be interpreted as a failed regression model; the data is not at all aligned with the regression plot as can be seen in both the schematic and our example.

Now we will define the last and most problematic output of a graph- The regression graph that doesn't contain a proper line. We added a schematic (*Figure 1.c:Plot A*) as reference to explain how a failed regression plot looks like in reality. Instead of the black linear line, the regression line is the blue line, whereas the predicted attribute is the orange line. We therefore added our own failed model's example, to help you observe the following:

1. There is no data trend
2. Most, if not all, molecules are outliers
3. This points to two things:
   → The data is not related (little or no correlation)
   → The descriptors contain extreme values (extreme negatives or extreme positives)

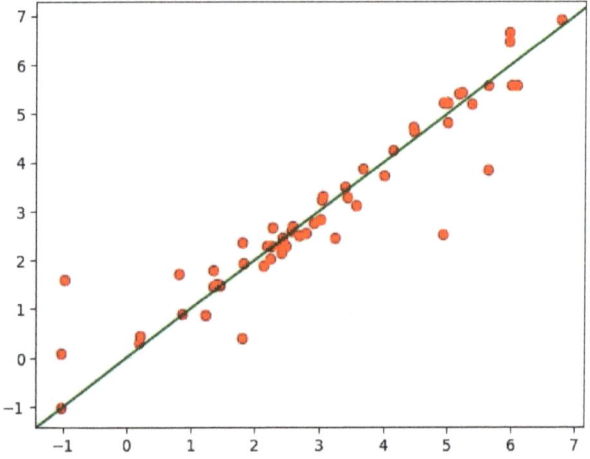

Figure 1.d: The figure shows a good regression plot; observe how there are outliers present and yet the data has a trend to follow. The R2 validation score for this was 0.871.

The above plot (*Figure1.d*) is the correct plot that we finally had received as an output after following proper data selection and descriptor selection rules. This is how a fitted plot ideally looks like.

Now let's talk about how non-linear or collinear graphs define the dataset:
1. For feature selection and significance
2. For Fit Plots

Type 1: Collinear model defining number of significant attributes in the descriptor set:

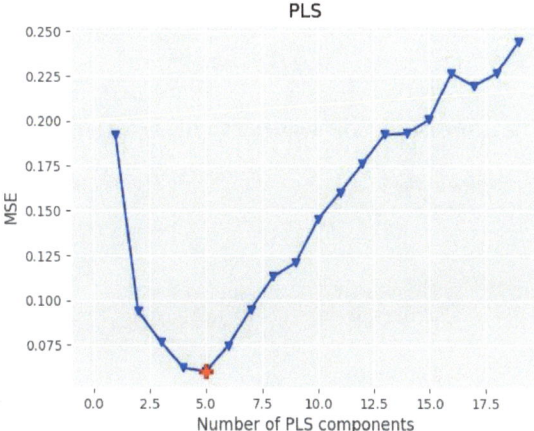

Figure 1.1: How a collinear model does feature selection as indicated by the red arrow, letting us know that there are 5 features that define the data perfectly; based on MSE (Mean Square Error).

Type 2: How a normal fit plot for a collinear model looks like:

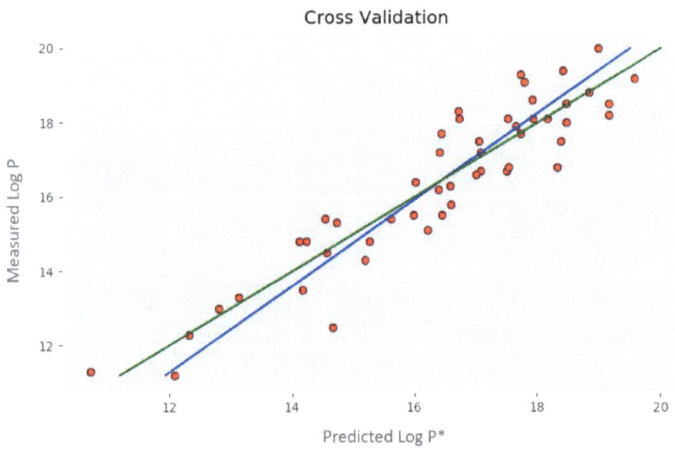

Figure 1.2: The figure shows how a normal **Partial Least Square Regression** fit plot looks like, if the data is fairly significant

Output 2:

Heat Maps: A significant feature selection method based on color coordinated scoring.

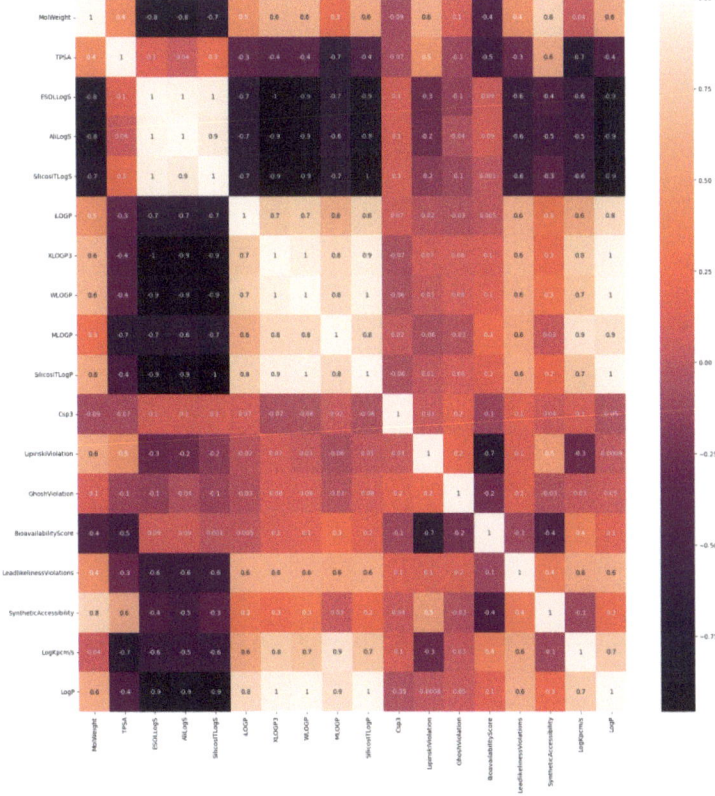

Figure 2.1: How Heat plot defines collinearity and redundancy in data.

A heat plot is actually used to define each descriptor's correlation with the dependant descriptor, in our case Log P, can be any other dependant variable and other cases. What itdoes is, it gives a colour scheme as well as a score, designating positive as well as negative correlation. Now here are a few things one needs to remember about correlation scores:

1. A high correlation can be both in the positive as well as the negative aspect.
2. Scores near 0.5 or -0.5 are insignificant for example, in Figure 2.1 the desctiptors that have yielded a score of 0.1, 0.2, 0.3, 0.4, 0.5, 0.6, 0.7 and -0.1, -0.2, -0.3, -0.4, -0.5, -0.6, -0.7. These value can be discarded right away from the list of significant descriptors.
3. A score of 1 means there is a perfect correlation where as scores of 0.9 are mostly accepted as significant descriptors.

Just like this, collinearity can also be defined from a heat map, by paying heed to the nature of those descriptors and their scores. If you see our heat map, there are several descriptors with the LogP preceeding them, thereby definingthem in a way as permeability descriptors.

Now, here is the main confusion that most people face, is a permeability study not needed? No, it is needed, but only a permeability study, if not completely novel in some aspect, can't hold up a research statement even with ample background research.

Therefore, keeping all this in mind, let us talk about the last part of the output.

Output 3:

Score: A way to understand the probable trend of the data.

There are two kinds of validation in a QSAR study:

1. R^2 validation
2. Rm^2 validation

Both of these validation matrices are cross-validations that can be performed 10 fold to - 50 fold cyclesto check the data over and over again, step by step.

Now here is another thing that these validations depend on- the error and scaling.

Every validation is done after scaling the data and once before scaling the data.

Scores are visualized in the following way:

Format of output:	Normal scores:	Good scores:	Bad scores:
R2 calibrated: 0.xxxx R2 CV: 0.xxxx MSE Calibrated: 0.xxxx MSE CV: 0.xxxx	R2 calibrated: 0.88234543 R2 CV: 0.8799652 MSE Calibrated: 0.0054 MSE CV: 0.0078	R2 calibrated: 0.998546 R2 CV: 0.997453 MSE Calibrated: 0.0026 MSE CV: 0.0031	R2 calibrated: -12.3465778 R2 CV: -11.3443778 MSE Calibrated: 0.890 MSE CV: 1.688
Without/ After Scaling: Rm^2: 0. xxxxxxx Reverse Rm^2:0. xxxxxxx Average Rm^2:0. xxxxxxx Delta Rm^2: 0. xxxxxxxx	Without Scaling: Rm^2: 0.727854 Reverse Rm^2: 0.7147788 Average Rm^2: 0.7129633 Delta Rm^2: 0.00853 After Scaling: Rm^2: 0.7139323 Reverse Rm^2: 0.70100044 Average Rm^2: 0.701119639 Delta Rm^2: 0.00884	Without Scaling: Rm^2: 0.9899311 Reverse Rm^2: 0.984755 Average Rm^2: 0.9873431 Delta Rm^2: 0.005176032 After Scaling: Rm^2: 0.9898217 Reverse Rm^2: 0.9843565 Average Rm^2: 0.9870891 Delta Rm^2: 0.005465167	Without Scaling: Rm^2: 0.07231479 Reverse Rm^2: 0.06847788 Average Rm^2: 0.07039633 Delta Rm^2: 0.003836904 After Scaling: Rm^2: 0.05339323 Reverse Rm^2: -0.03100044 Average Rm^2: 0.01119639 Delta Rm^2: 0.08439367
R squared: 0. xxxxxx Adjusted R squared:0. xxx	R squared: 0.8927548 Adjusted R squared: 0.8920562	R squared: 0.9927584 Adjusted R squared: 0.9920342	R squared: 0.2520198 Adjusted R squared: 0.1756953
Average R: 0. xxxxxx Average R^2: 0. xxxxxx Average Q^2: 0. xxxxx cRp^2: 0. xxxx	Average R: 0.2651342 Average R^2: 0.08431973 Average Q^2: -0.1998938 cRp^2: **0.7534394**	Average R: 0.2776342 Average R^2: 0.08439137 Average Q^2: -0.1955338 cRp^2: **0.9534394**	Average R: 0.2819334 Average R^2: 0.08242916 Average Q^2: NA cRp^2: **0.208523**

Table 1: The table describes how scores are given and their example formats, along with normal, good and bad score examples.

Here, all scores in the table are the scores that we got as the model went from wrong to right and then normal, yes in that order

Let's discuss the meaning of these scores:

1. Validation scores tell us how capable a model is of predicting the property of an unknown molecule, with differing biological activity values.

2. Finding validation score based on error, here mean square error: R square CV is calculated as the ratio of the mean squared error (MSE) to the mean of the dependent variable (Y).
3. cRp^2 is known as the Y scrambling parameter and is a value given for probable similarity between predicted and measured Y or Dependant parameters; this value is necessary during the model validation through Y-randomization.
4. Adjusted R square or Rm square (R^2 or Rm^2) is a score that gives the corrected goodness of fit calculation of the model's validation.
5. Delta Rm^2 is a value of confidence, or the confidence level of a model, commonly known as the p-value, the lesser the value the stronger the model.
6. Finally we come to calibrated or scaled values, these terms mean the same, and define the data, that is closely related. It is to be noted that if the score of a model increases or decreases too much after calibration or scaling, then there are some errors in the given model, which you would notice in the wrong model score in the table *(table 1)*.
7. The last but not least, what does 'NA' signify, as a score? It means that there is an error in the data set, literally, please go back and check!

As we can understand now, there are various outputs and various ways to decipher those outputs. This is why understanding the output of a model is so critical. Now, we would have explained exactly how these scores matter, but that is for another chapter, completely dedicated to that. So stay tuned and move on to the next chapter.

How the Data and Model validation go Hand in Hand?

We understood the importance of good data and proper choice of model in the prior chapters. But didn'twe just discuss that good data can also give low scores? Didn't we say that good data can sometimes be designatedas inadequate or even bad data during model validation?

This is why we emphasized on model selection since just like life if we match the right people with wrong timing, situations can go out of hand; a wrong model with right data or right model with dissimilar data can have acatastrophiceffect.

The mathematical understanding of QSAR, unlike other methods, isvery much real time based, and that is why it is employed over and over again to understand environmental toxicity or dissimilarities. At the same time, timing and situation too play a strong role during data selection before a model validates the data.

Situations can imply a variety of ideas like:

1. Optical situations → levo and dextro molecules
2. Regional dimilarities in molecular behaviour
3. Seasonal similarities
4. Disease- specific similarities

5. Similarities in Reactivity profile

But one aspect is common about all of these situations: their biological activity. Their biological activity is what makes them similar and what does that imply then... their structural similarity.

Unlike numerical similarity measures, substructure-based methods give a simple binary (Yes/No, True/False etc.) answer to whether two compounds are similar. One common approach is to cluster compound datasets based on hierarchical scaffolds that represent core structures. Another way to measure substructure-based similarity is by finding the maximum common substructure of compounds, although this method is more useful for smaller datasets. On the other hand, similarities can also be determined on a larger scale using the matched molecular pair formalism (MMP), which operates similar to scaffold analysis.

An MMP refers to two compounds that only differ in a modification at one location. This means that both compounds have a shared core structure, and the modification can be described as swapping out a pair of substructures, which is called a chemical transformation. Generating MMPs using algorithms is very efficient. By restricting the size of transformations, it's possible to generate pairs of analogs as MMPs. Using a combination of MMP search and network analysis, allows the QSAR practitioner for a systematic extraction of analog series from large compound sets and enables further exploration through SAR investigation and QSAR modeling.

Expanding the scope of QSAR analysis involves moving away from the traditional approach and considering compounds with diverse structures that don't share common features. In these cases, there are significant structural differences between active compounds that can't be modeled using linear approaches like "scaffold hopping." This complexity makes it challenging to accurately predict bioactivity. Non-linear SAR models are needed to understand the relationship between structure and potency in both similar and dissimilar analogs. However, Machine learning techniques, play a crucial role in handling this level of complexity beyond what classical linear regression QSAR methods can handle.

In summary, the selection of molecular descriptors and evaluation of molecular similarity are crucial factors in quantitative structure-activity relationship studies. It is important to note that comparing representations of objects, determining their similarity metrics, and understanding the relationships between object properties are relevant across various research fields. Paracelsus' principle "similars cure similars" can be considered a fundamental way of thinking that applies not only to toxicology but also to cheminformatics and diverse scientific disciplines. This principle underlies the effectiveness and versatility of approaches used in cheminformatics, as illustrated throughout this discussion.

Deep understanding of R^2 and Rm^2 model validation scores- A low score isn't always bad?

Don't define your world in Black or White, because there is so much hiding in Grey

Just like the quote states, there is no black or white, there is no answer that is completely wrong or completely right, there is no score that is wrong or right.

It isthe technique that decide that for us, the way we decipher our path and clear the path to travel towards the end goal that defines our actions.

QSAR modeling, even though it is just a linear method of modeling, is one of the toughest methods when it comes to defining the data. This technique not only defines data but even validates their biological synergy to explain the importance of the molecules involved.

This method's strongest validation points are- R^2 validation and the Rm^2 validation.

To truly understand the strength of a model, it is crucial to examine its R2 value. This numerical representation showcases how closely aligned the predicted values are with the observed outcome values - essentially measuring their correlation squared. For optimal performance, aiming for a higher adjusted R2 should be our top priority. Rm^2 on the other hand considers the precise distinction between the observed and predicted response data, disregarding the average of training set. This makes it a more rigorous criterion for evaluating the predictability of a model compared to conventional validation parameters.

Therefore, when these validations parameters give a score, there can be several ways to interpret them:

1. A low score: A low score means 3 aspects:

    → The molecules are highly toxic

    →The data set is not fitted with the model

    →The model is a failed model

2. A high score:

    → The data fits the model

    → The data portrays structural and functional similarity

    → The model has been properly validated

But the flip side of a high score?

   a. The model can be over-fitted
   b. The model can be collinear
   c. The model can have too less values to predict anything at all, and the score can be a fluke.

Whereas, when it comes to lower scores, we can define the perfect toxic variable and molecule from that model. This might help in bioremediation or simple filtration which can potentially remove this or rather these molecules from a place/environment...

Not just this, low scores create contradictions in studies, while one might take a low score as a failed model 99 percent of times, that last 1 percent can prove to be valuable and viable information for a completely different start of a novel study.

Therefore, never underestimate a low score, yes re-check and re-evaluate but never disregard. Let us bourne in mind that a failed experiment can act as a seed to a bigger and better plan in the long run, and teach us all the paths not to tread, to arrive at the end goal.

Concluding Statements:

Author Contribution:

All authors have contributed to this work equally.

External help and funding:

No external funding or help was required or received for this study.

Applications involved:

The Applications involved in this study were:

1. SWISSADME (http://www.swissadme.ch/index.php)

2. Protox-Tox II (https://tox-new.charite.de/protox_II/)

3. R Lab (installed in local)

4. Google Colaboratory (https://colab.research.google.com/)

5. PaDEL (http://www.scbdd.com/padel_desc/index/)

6. PubChem (https://pubchem.ncbi.nlm.nih.gov/)

7. PubMed (https://pubmed.ncbi.nlm.nih.gov/)

8. PDB (https://www.rcsb.org/)

Glossary:

Key terms		Meaning
1.	QSAR	A cheminformatic study of the biological activity and toxicity of a set of molecules with the help of descriptive attributes.
2.	Data set	A group of molecules, preferably more than 55
3.	Data filtration	A group of molecules formed after they have been sorted according to structural similarity.
4.	Descriptor	A term used to define attributes that describe molecules. (Like solubility, permeability, etc.)
5.	Descriptor filtration	Sorting of attributes in a way that there is no redundancy(repetition) or collinearity in data.
6.	Model	A mathematical program that using logic and specific rules defines a data set.
7.	Score	A value that defines the model's power to predict an outcome.
8.	Graphs	A pictorial representation of data trends
9.	Regression Plot	A graph generated to prove similarity between the dependent entity and the predicted dependent entity.
10.	Heat maps	Pictorial colour-coded scores assigned on the correlation between two descriptors.
11.	Biological Activity	A molecule's nature/ how it will behave in a living environment.
12.	Validation	A process to define the correctness of a model.
13.	Validation matrix	A scoring matrix that scores the probabilistic capabilities of a model.
14.	Dependant variable	The main entity that defines the data and on which the other variables are dependent, like, reactivity, permeability,

		etc.
15.	Independent variable	Variables that define a molecule, can be any variable, physical, biological, and so on. These variables do not have other variables dependent on them.
16.	Dependency	Relation between two descriptors and if they are directly or indirectly proportional.
17.	Linear data	Data that is not a sub-definition of another variable/ or related to another variable.
18.	Collinear data	Data that are closely related, and if any are missing, other similar variables can define the same attribute.

Acknowledgement

We would like to acknowledge Arushi Das, a fellow friend and co-worker for her insights on the work and we would especially like to thank Dr. Kunal Roy for his inspiring work that originally motivated us to execute this work.

www.ingramcontent.com/pod-product-compliance
Lightning Source LLC
Chambersburg PA
CBHW040324220526
45473CB00009B/2557